FLORA OF TROPICAL EAST AFRICA

LECYTHIDACEAE

G. R. W. Sangai

(East African Herbarium)

Trees or shrubs. Leaves simple, alternate, exstipulate or stipules small. Flowers usually rather large and showy, regular or irregular, hermaphrodite. Calyx 2–6-lobed or circumscissile, if the lobes free then valvate or slightly imbricate. Petals 4–6 or lacking, free or connate at the base. Stamens numerous, in several rows, free or mostly basally united into a short tube, the inner filaments shorter and often without anthers; anthers basifixed or rarely adnate, opening by slits. Ovary inferior or half-inferior, 1–6(–many)-locular; ovules 1–many in each locule, axile or pendulous from near the apex; style simple or shortly branched at the apex. Fruit woody, fibrous or fleshy, indehiscent or operculate. Seeds without or with thin endosperm; embryo straight or curved, sometimes solid and undifferentiated, but in many genera with normal cotyledons.

A moderate-sized family, formerly included in the *Myrtaceae*, of 22 genera, only two of which are recorded from the Flora area, namely *Barringtonia* J. R. & G. Forst. and *Foetidia* Lam. These and several others are often considered to belong to a distinct family, *Barringtoniaceae* (see Knuth in E.P. IV. 219 and 219a, Barringtoniaceae and Lecythidaceae, 1939). Very recently *Foetidia* has been placed in a separate family (see Airy Shaw, Dict. Fl. Pl. & Ferns, ed. 7: 454 (1966)).

The cannon-ball tree, *Couroupita guianensis* Aubl. is in cultivation in Uganda (Entebbe Botanical Garden, see Dale, Introd. Trees Uganda: 25 (1953)), Tanganyika (Lushoto District, Amani, 28 Dec. 1939, *Greenway* 5911!) and in Kenya (Mombasa). It belongs to the *Lecythidaceae* sensu stricto.

Petals 4–5(–6), conspicuous; style simple, subulate at the tip; flowers in racemes or clusters; leaves not involute   1. **Barringtonia**

Petals absent; style with 3–5 branches; flowers axillary, solitary or arranged in cymes; leaves involute.   .   2. **Foetidia**

1. BARRINGTONIA

J. R. & G. Forst., Char. Gen. Pl.: 75, t. 38 (1776); Payens in Blumea 15: 157–263 (1967) [very full monographic treatment], *nom. conserv.*

Trees and shrubs. Leaves mostly tufted towards the ends of the branches, entire or slightly crenate-serrate; stipules very small, deciduous. Flowers mostly scented and often opening at night, in erect or pendulous, terminal or lateral racemes; bracts and bracteoles small, soon deciduous. Calyx-tube not or hardly produced above the ovary; limb of 3–5 free lobes or entirely closed before anthesis, then 2–5-lobed or circumscissile, persistent. Petals 4, rarely 5–6, adhering to the staminal tube at the base. Stamens connate into a short tube at the base; inner ones shorter, without anthers. Ovary 2–4(–5)-locular; ovules 2–many in each locule, mostly pendulous from the apex of the inner angle. Fruit with fleshy-fibrous exocarp and a woody-fibrous endocarp, indehiscent, usually 1-seeded by abortion. Seeds spindle-shaped or trigonous. Embryo undifferentiated, without cotyledons; no albumen.

A genus of 39 species, only two of which are recorded from East Africa, widely distributed from East Africa, tropical Asia, Malaya to Australia and Polynesia, often near the sea.

Racemes erect, stout, terminal; leaves entire, sessile, obovate; fruits broadly pyramidal, obtuse, 4-angular, up to 12 cm. long and wide .   .   .   .   .   1. *B. asiatica*
Racemes pendulous, slender, axillary; leaves slightly crenulate, very shortly petiolate, oblong-obovate to elongate-oblanceolate; fruit ± ellipsoid or conic, obtusely 4-angular below, 3–7(–9) cm. long, 2–4(–5·5) cm. wide .   .   .   .   2. *B. racemosa*

1. **B. asiatica** (*L.*) *Kurz*, Rep. Pegu., App. A: lxv, App. B: 52 in key (1875) & in Journ. As. Soc. Bengal 45 (2): 131 (1879); Knuth in E.P. IV. 219: 10 (1939); T.T.C.L.: 263 (1949); Backer & Bakh. f., Fl. Java 1: 352 (1963); Payens in Blumea 15: 184 (1967). Type: Java, Prinsen I., *Osbeck* (LINN, specimen 675.2, holo., S, iso.)

Tree or shrub 7–30 m. tall, with grey or brown, rough or smooth scaly thick bark; trunk sometimes buttressed; young branches with large leaf-scars. Leaves sessile, obovate or oblong-lanceolate to oblong-obovate, (5–)16–40(–56) cm. long, (2–)10–18(–25) cm. wide, obtuse, broadly rounded, slightly emarginate or even somewhat acuminate at the apex, narrowed to a ± rounded-truncate base, coriaceous, glabrous, often shining, entire; larger leaves often alternating with smaller ones. Racemes terminal, erect, 2–30 cm. long, 3–20-flowered; pedicels 4–6(–10) cm. long. Calyx 2-lobed, the lobes 3–5 cm. long, 2–3·8 cm. wide. Petals white, broadly elliptic, 5·5–8·5(–?10) cm. long, 2–4·5 cm. wide. Outer stamens numerous, 6–12(–15–?18) cm. long, mostly pink above. Ovary 4-angular, 4(–5)-locular, with 4(–5) ovules in each locule; style 9–13·5(–15–?17·5) cm. long. Fruit broadly pyramidal, 4-angular, 8·5–12 cm. long and wide; pericarp thick, glossy, the endocarp fleshy-fibrous. Seed oblong, 4–6 cm. long, 2·5–5 cm. wide, subtetragonous, tapering to an emarginate apex. Fig. 1/8.

ZANZIBAR. Zanzibar I., without precise locality (fruit only), *Barraud*!; Pemba I., W. coast, Verani, 18 Feb. 1929, *Greenway* 1468! & Chake Chake, 13 Aug. 1929, *Vaughan* 495!*
DISTR. **Z**; **P**; R. O. Williams (U.O.P.Z.: 141 (1949)) also records it from Wete in Pemba and Mbweni on Zanzibar I. where it is perhaps cultivated; Seychelles, Mauritius, Comores and Madagascar, thence from India (Coromandel coast) throughout tropical Asia to S. Vietnam, Malesia, Formosa, Queensland far into the Pacific; also introduced into Hawaii, West Indies, St. Helena and Guyana
HAB. Sandy or rocky beaches; sea-level

SYN. *Mammea asiatica* L., Sp. Pl.: 512 (1753)
    Barringtonia speciosa J. R. & G. Forst., Char. Gen. Pl.: 76, t. 38 (1776); C.B. Cl. in Fl. Brit. India 2: 507 (1879); U.O.P.Z.: 140 (1949). Type: Tahiti, *Forster* (BM, holo.)

NOTE. This is a characteristic drift fruit of the Indopacific region and fruits have been picked up on the East African mainland, e.g. Kenya, Kwale District, N. of Jadini, 12 Dec. 1959, *Greenway* in *E.A.H.* 11833! The species has also been collected from the Dar es Salaam Botanic Garden, 7 Oct. 1929, *Burtt Davy* 22324, presumably from a cultivated plant.

2. **B. racemosa** (*L.*) *Spreng.*, Syst. Veg. 3: 127 (1826); DC., Prodr. 3: 288 (1828); Harv. in Fl. Cap. 2: 523 (1862); Lawson in F.T.A. 2: 438 (1871); C.B. Cl. in Fl. Brit. India 2: 507 (1879); V.E. 3(2): 658, fig. 290 (1921); Knuth in E.P. IV. 219: 17 (1939); U.O.P.Z.: 139, fig. (1949); T.T.C.L.: 263 (1949); Perrier de la Bâthie in Fl. Madag. 149: 4, fig. 2/6–9 (1954); E.P.A.: 613 (1959); K.T.S.: 243, fig. 48 (1961); Backer & Bakh. f., Fl. Java 1: 353 (1963);

* In his notebook Vaughan gives " Chake Steps (Cult. ?) ".

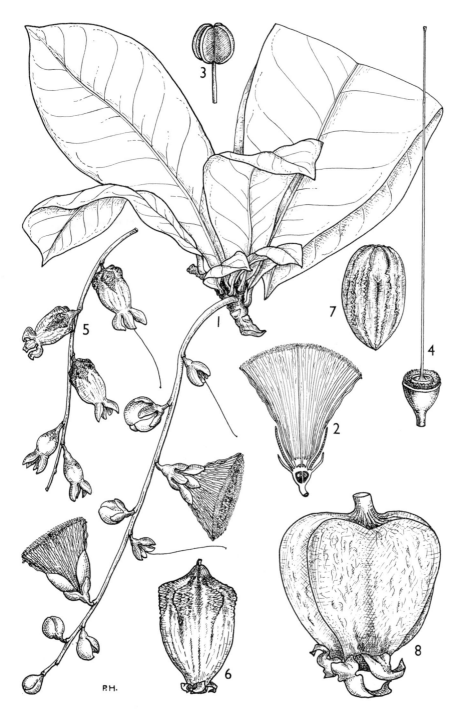

FIG. 1. *BARRINGTONIA RACEMOSA*—**1,** flowering branchlet, × ½; **2,** section of flower, × ⅔; **3,** anther, × 16; **4,** gynoecium, × 2; **5,** young infructescence, × ½; **6,** fruit, × ½; **7,** seed, × ⅔. *B. ASIATICA*— **8,** fruit, × ¼. 1–5, from *Verdcourt* 1749; 6, from *Tanner* 2654; 7, from *Paines & Reid*; 8, from *Barraud*.

Pacific Pl. Areas 2 (Blumea Suppl. 5), Map 164 (1966); Payens in Blumea 15:
192 (1967). Type: Ceylon, Herb. Hermann figs. 212, 213 & 339 (BM, syn.)

Tree 2–12(–27) m. tall, with grey and yellow, greyish or yellowish-brown
smooth or fissured bark, sometimes * with pneumatophores as in mangroves;
bark on young branches very tough; leaf-scars conspicuous; crown with
spreading branches. Leaves shortly petiolate; blades obovate-oblong, or
usually elliptic, to elongate-oblanceolate, (5–)15–36(–42) cm. long, (2–)6–
14(–16) cm. wide, acuminate at the apex, cuneate at the base, often rounded
at the extreme base, glabrous, crenulate; petiole 0·25–1·5 cm. long. Flowers
many in pendulous racemes 20–70(–100) cm. long; pedicels 0·3–1·6(–2·5) cm.
long. Calyx green, mostly flushed pink or deep purple, splitting into (2–)3–
4(–5) lobes, 0·75–1·5 cm. long, 0·5–1·25 cm. wide. Petals white, often tinged
pink outside, elliptic, 1·2–3 cm. long, 0·5–1·5 cm. wide, adnate to the staminal
tube. Stamens white or pinkish tinged, numerous, in several rows, 2·2–
3·7(–5) cm. long. Ovary ± globular, (2–)3–4(–6, fide Kirk)-locular; ovules
2–3 in each locule; style red, (2–)3–6·2 cm. long. Fruit ± ellipsoid or conic,
round in section or sometimes distinctly angular, 3–7(–9) cm. long, 2–4(–5·5)
cm. wide, scarcely glossy, mostly dull and rather rough; endocarp fleshy-
fibrous. Seed ovoid or subtetragonous, 2–4 cm. long, 1–2·5 cm. wide.
Fig. 1/1–7.

KENYA. Kwale, Aug. 1929, *R. M. Graham* in *F.D.* 1969!; Kilifi District: Rabai, Aug.
1937, *V. G. L. van Someren*! & Ribe, *Wakefield*!
TANGANYIKA. Lushoto District: Korogwe, Jan. 1903, *Zimmermann* 176!; Tanga
District: Magunga Estate, 13 Feb. 1953, *Faulkner* 1140! & Sigi valley, 13 km. below
Amani, 29 Dec. 1956, *Verdcourt* 1749!; Uzaramo District: Kisiju, Sept. 1953, *Semsei*
1375!
ZANZIBAR. Zanzibar I., Sept. 1873, *Kirk*! & *Lyne* 73!; Pemba I., Shengejuu–Weani,
19 Feb. 1929, *Greenway* 1507!
DISTR. **K**7; **T**3, 6, 8; **Z**; **P**; Somali Republic, Mozambique, Natal, Seychelles, Comores,
Madagascar, thence extending to S. Asiatic coast, Malesia, tropical Australia and into
the Pacific
HAB. In forest by rivers and streams in coastal belt and some short way inland, also
often between the mangroves and bordering grassland or bush with roots often in the
water; 0–450(–750) m.

SYN. *Eugenia racemosa* L., Sp. Pl.: 471 (1753)

NOTE. The seeds are dispersed by water.

2. **FOETIDIA**

Lam., Encyl. 2: 457 (1788) & Tabl., t. 419 (1794)

Faetidia Juss., Gen.: 325 (1789); Lam., Tabl. 2: 529 (1819)

Small to medium-sized glabrous trees. Leaves alternate, borne towards the
tips of the branches, persistent or deciduous, petiolate, coriaceous, entire,
pinnately nerved, involute when young. Flowers axillary, solitary or in small
cymes, 3–5-merous, apetalous. Receptacle top-shaped. Sepals 3–5, coriaceous, valvate or reduplicate-valvate, persistent, sometimes enlarging on the
fruit. Stamens numerous, in many series, sometimes in 4 groups opposite
the sepals; filaments free, unequal, often bent in bud; anthers elliptic or
oblong. Ovary inferior, 3–5-locular, the locules equal in number to the sepals
and alternating with them; style slender, shortly divided into 3–5 branches
at the top; ovules 15–20 in each locule, horizontal or oblique, anatropous.
Fruit cone- or top-shaped, very tough or woody, round in section or with
3–5 wings or angles alternating with the sepals, 1–4-locular, each locule

* Mentioned in some very full and interesting notes by Sir John Kirk preserved in
the herbarium at Kew.

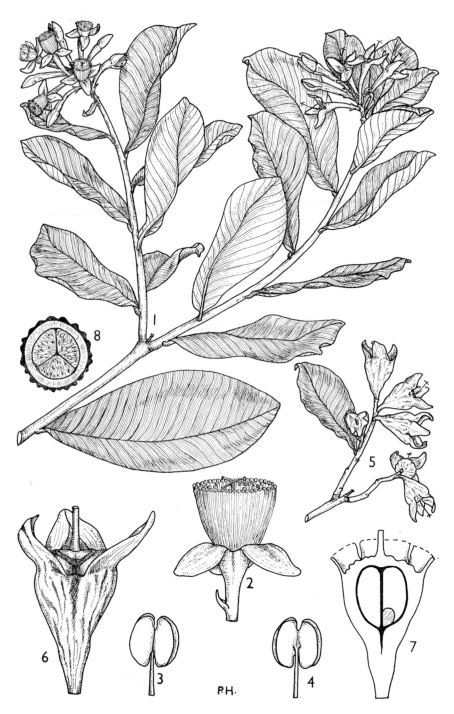

FIG. 2. *FOETIDIA OBLIQUA*—**1**, flowering branch, × ⅔; **2**, flower, × 2; **3**, anther, front view, × 20; **4**, same, back view, × 20; **5**, fruiting branchlet, × ⅔; **6**, fruit, × 2; **7**, longitudinal section of fruit, × 2; **8**, transverse section of same, × 2. 1–4, from *Scott Elliot* 2620; 5–8, from *Decary* 10120.

1-seeded. Seeds small, elongate or subglobose; albumen thin and fleshy; embryo narrow, slightly curved, with a long radicle.

A small genus of 5 species, 4 confined to Madagascar (save for the occurrence mentioned beneath) and one endemic to the Mascarene Is.

F. obliqua *Blume*, Mus. Bot. Lugd.-Batavum 1: 145 (1850); Grandidier, Hist. Nat. Pl. Madag., Atlas, t. 343 (1895); Knuth in E.P. IV. 219: 63 (1939); Perrier de la Bâthie in Fl. Madag. 149: 7, t. 3/6–10 (1954). Type: Madagascar, collector not stated (P, holo.)

Small glabrous evergreen tree or large shrub. Leaves sessile or with very short petiole; blades elliptic to oblong-obovate, 3·5–11·5 cm. long, 1·2–5 cm. wide, rounded, acute or slightly acuminate at the apex, mostly cuneate at the base, coriaceous, the midrib slightly asymmetric, dividing the leaf into unequal parts; lateral nerves fine, very oblique and crowded, rendering the species recognizable at a glance. Flowers 6–25, often condensed at the tip of a branch; pedicels 0·3–1·2 cm. long; bracts linear-lanceolate, 0·3–1·3 cm. long, 1–2 mm. wide; bracteoles lanceolate, 1–3 mm. long. Receptacle obconic, 7 mm. long, 5 mm. wide, strongly striate. Sepals broadly rounded-deltoid to lanceolate, 0·5–1 cm. long, 4–7(–8) mm. wide. Stamens numerous, about equalling the sepals; anthers 0·4 mm. long. Style ± 1·2 cm. long, with terminal branches 1·6–1·7 mm. long. Fruit red, obconic, 1·2–1·4 cm. long, 0·8–1·1 cm. wide, strongly lined lengthways, with persistent sepals crowning the fruit and the same length as at anthesis but mostly wider. Fig. 2.

ZANZIBAR. Pemba I., Chake Creek, 15 Sept. 1929, *Vaughan* 651!
DISTR. P; eastern part of Madagascar from Antalaha to Fort Dauphin
HAB. Presumably in mangrove swamps; sea-level

NOTE. Further information on the status of this plant in the Flora area is required. There is no reason why it should not be native, but whether a recent arrival or a relict of a once wider distribution is unknown.

INDEX TO LECYTHIDACEAE